Miracle

Mudra

Girish Kamanuri

Miracle Mudra

Girish Kamanuri

Miracle Mudra

Index

Introduction		~ 8
1.	Mudra	~ 26
	➤ Agni (fire)	~ 35
	➤ Vayu (Air)	~ 45
	➤ Aakasha (space)	~ 54
	➤ Prithvi (earth)	~ 61
	➤ Jal (water)	~ 70
2.	Major Mudra	~ 78
3	Human life and Mudra	~ 81
4	How to use mudra	~ 87
5	Knowledge through mudra	~ 91
6	Positive energy through mudra	~96
7	Human Brian and mudra	~103
8	Mudra a Major direction	~111
9	How to unlock Mudra	~115
10	MIRACLE MUDRA	~120

Girish Kamanuri

Miracle Mudra

Girish Kamanuri

Dedication

This book is dedicated for those who are unaware of the supernatural powers which are present in their own hands and for those who are seeking positive energy or knowledge.

Miracle Mudra

Girish Kamanuri

Miracle Mudra

Acknowledgements

I thank my Guru (teachers) for giving me such a wonderful knowledge in my life towards education, medicine and science derivation.

Girish Kamanuri

Miracle Mudra

Girish Kamanuri

Introduction

Mudra, consider it as magical remedy for every living human being to study past, present and future. This book is completely based on the true aspects of using mudra towards medicinal and scientific derivations.

Life is a mystery and to unlock the mystery one must practice mudra towards science derivations. A mudra is beneficial only when there is space in the brain to attain maximum storing capacity. This means that the brain is unlocked until the human uses mudra to unlock major derivation as PATIENCE.

Patience is the major key and required to be practiced along with mudra practice. A mudra is activated only when there is movement practiced by the individual to know all five fingers and their integrity in gaining knowledge and to derive science. A mudra can be a magical spell for unlocking brain functioning in human life.

Miracle mudra book is based on the true story on unlocking human brain towards derivation of science through materialistic as well as through spiritual aspects of knowledge.

To unlock the mudra in human life, one must practice patience along with natyashastra to understand the knowledge derived through mudra. It is definitely not easy to understand the knowledge within the brain derived by mudra practice. A mudra has got magical powers towards healing, educate and transport.

Complete book is based on how the mudra technique can be bought into someone's life with experiences made through series of experiments. One must know that this book do not just reveal the answers in the brain but also brings positive note on the purpose of this life while practicing or during the practice.

It's easy to hold the mudra for few minutes but hard to sit for more than 10 or 15 minutes in the beginning. Why is this happening in the beginning? Is it natural or the technique is wrong? I believe every first individual will experience these questions in their mind before even the mudra is working for them.

A mudra cannot be used as just an experimental way, but I believe one must hold the mudra for ENLIGHTENMENT. Energy and mudra are having beautiful communications, where energy is stored while doing mudra practice. This means the energy is being stored and understood without even moving an inch. How is this possible? (Laughs) it is possible according to me. When you are moving, the energy can be felt through the emotions of pain and while sitting and holding mudra you do not just feel the energy but you can save the energy within oneself.

As a dance scientist I have always believed in the movement practice from the beginning, when I was introduced to this mudra technique, I was left SPEECHLESS. A dance can bring the energy as well as happiness and a mudra can bring the same energy and happiness without even making a single move. How is it possible (you might be thinking)?

If you dance for a while the energy can be felt within just 10 minutes of practice itself but in case mudra alone, one must practice at least for

Miracle Mudra

10 hours to feel the energy surrounded while doing meditation. The major answer to reduce the time for realization point of mudra technique is been shared in the last pages of this book. If I announce the technique in just few words then it's hard to understand and to feel the energy within. I do not want anyone in this world to use the mudra technique for wrong purposes. The one who uses the mudra for wrong purposes I believe they never understood the energy through mudra.

Miracle mudra book is based on not just how to use the technique but also explains about the ways of reaching out to energy for good purposes through mudra practice.

Usually I have seen my students saying that it's hard to close the eyes while doing mudra and hard to concentrate while doing meditation through mudra. There is nothing wrong in this, I believe as a scientist the emotions comes in front of your eyes as series of images while doing the mudra meditational practice. This is natural but prolong practice of meditation through mudra gives you strength to close your eyes calmly without any images coming in front as well. One must educate their brain (monkey mind) towards peaceful meditation through mudra. If the images are troubling you continuously while doing the mudra practice or meditation then you have imbalance in your brain functioning. It takes at least 6 weeks for one to understand how to use and hold mudra while dong meditation.

You might be wondering which is the miracle mudra, which brings peace, harmony, wealth, happiness, stress relief? (Laughing) As a dance scientist I said earlier that PATIENCE is and will be major key. One who reds this book will receive an amazing answer for the above question in end. This is because I want to explain the mudra technique in very strong manner, so that one who practices must feel the energy within 10 minutes itself and not in 10 hours.

People who want to read this book might be eagerly waiting to know which the most amazing miracle mudra is and how it works to solve all the problems or worries in life. This book make will make sure that knowledge has no boundary but the technique has exact boundary.

To understand miracle mudra one needs to control their brain functioning through natya (dance), Swara (sing), Music (taala). Once you have control on your brain functioning then the mudra technique become more valuable in terms of gaining knowledge and understanding the energies surrounded by.

A mudra can be life saving. Is it true or it's just a saying? I believe as a dance scientist that every mudra has unique powers to describe and to derive. To derive such powers one must realize their purpose of living. This means that one must surrender to the nature to read the KNOWLEDGE.

There is millions of mudra according to me and each one of those has equivalent powers to describe. Then which is the miracle mudra to derive Science, medicine and education? (Again Laughs) there are millions of people out there in this world who is still seeking the miracle mudra for their benefit. Most of them have been practicing mudra for an hour and they usually change it for their convenience.

Is this way of using mudra is correct? After one hour change the mudra and see what brain gives you answer? (Smiles) According to me every mudra has got magical powers and to obtain or to understand such powers one need study meditation for prolonged time of period, probably a month or a year. So you might be thinking who is going to sit for ten days continuously for meditation using mudra right? As I mentioned earlier, at first patience is the key to achieve before you achieve mudra.

Miracle Mudra

The entire above introduction might be confusing you with respect to the title Miracle Mudra. Trust me the magic of mudra is amazing as it is practiced with prolonged time with patience.

Many of the world's best teachers use Mudra technique to educate themselves. One who practices mudra will get benefits from the entire universe towards education, transformation and healing.

The technique of mudra is exquisite in terms of knowledge. I have seen many people wanting to squeeze their fingers to seek the answer very quickly. This kind of technique always gives negative aspect of knowledge and perception. Why?

Miracle Mudra

Try it out and see. To hold the mudra one must calm their mind as well as their body functioning, like to hold the mudra just through a TOUCH and not by pressing the fingers too hard to seek the benefits in faster rate.

Emotions are amazing if one expressed with feelings, otherwise every emotion will look like an acting way. An act must be so pure that one must feel the emotions to enjoy the perception. A direction must be novel way to spread smile and happiness.

A wise man once said that nothing is permanent but knowledge is one thing which is not just permanent but also spreads throughout the universe with experiences. A mudra has the same magical powers which can enlighten one's mind towards transformation.

I remember when I started using mudra for more knowledge I was confused in the beginning but have found out the ways to use mudra technique to enlighten my brain functioning into next level.

Knowledge cannot be in the same way expressed in the beginning it expands just the way, your brain wants to. A brain is designed to achieve the goal in many ways just like that every brain has portions or compartments which can be opened or unblocked with particular technique.

Miracle Mudra

There are seven chakras in human life and one must know that mudra can unblock one of the chakras in human life. This means very clear that only mudra technique alone cannot unblock every chakra in human life but by using the other techniques as well like, Natya (dance), Swara (sing), Meditation (tapassya) and many more. As a dance scientist I expanded more of my knowledge by practicing dance (natya) as well as practicing mudra along with postures to unblock various portions of my brain.

Miracle mudra is a beautiful explanation of unblocking the chakra perfectly. Every mudra is amazing and I believe if you are interested in knowing which miracle mudra gives every emotion to lead life, then you must wait till the end to know the answer derived for better living purpose.

A life without mudra is like a life without proper oxygen. Mudra and its significances are wide spread but the technique of using the mudra is not spread everywhere. It's true that many millions in this world practices mudra but most of them do not get the results. Why is that they are not getting the results? Is it due to the way they hold mudra or anything else? As a dance scientist and a mudra expert believe that most of the people who use mudra have no idea about which mudra is been designed for them or which purpose. A mudra chooses the right person or the person chooses the right mudra is the major answer in this book.

Holding mudra technique is as easy as having oxygen every day, but hard to choose the perfect mudra. Life is all about choosing the direction to lead life towards beautiful purposes. As mentioned in previous line, we need to choose the right mudra for ourselves to lead life and to unblock one of the chakra in human body.

It is easy to change the mudra and shift to another mudra but hard to receive benefits in this manner. Just it's easy to receive hate from everyone in kaliyuga but hard to get respect and love. As a dance scientist I have studied moral life rules to understand the medicine and science derivation. Imagine if I wouldn't have practiced moral ethics in my life for 20 years then I wouldn't be in this stage to spread free medicinal deviation through NGO and to derive science through natya (dance), mudra along with yogic postures.

Mudra

There are many different mudra practiced by human being. Every finger has its own identity. The thumb finger indicates FIRE. Complete mudra technique is based on how do we use fire. From the ancient times in India people practice Homa or Havana with Agni (fire). This indicates that one must bow down to the main source of energy which is FIRE (SUN) for every prayer you do or practice.

Fire Finger carries energy to all the other fingers which stimulates particular power. After the Fire(agni) finger comes is VAYU(air), continued with middle finger which is AKAASH(SPACE), moving to next

Miracle Mudra

finger which is PRITHVI(EARTH), and finally with last finger JAL(water). The complete above mentioned powers of each finger is shown in the figure below.

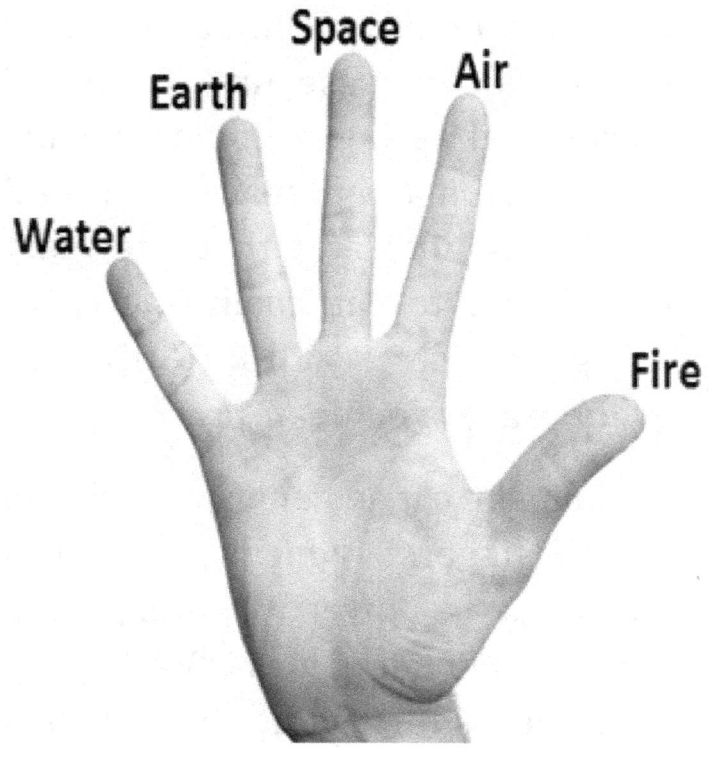

Miracle Mudra

Each one of must know the importance of each finger before we practice mudra to gain many significances. Each finger is essential in human life, so we need to achieve mudra technique by understanding the importance of each finger first and then to start with mudra practice for medicinal or scientific derivation.

To begin with mudra practice, one must know that mudra unlocks one of the chakras out of seven chakras in human life. This means very clear that in kaliyug if you are smart enough then you will adopt not only mudra technique but also adopt other techniques like Natya (dance),

Miracle Mudra

Swara (sing), Taala (music) and a good character.

The above paragraph mentions about the every chakra to activate then why one must adopt mudra technique for knowledge or to gain acceptance? In olden time we have seen the JAINISM involved in mudra technique for lifetime to educate, transport and heal. Then how this is possible? It is possible if the individual dedicated the entire life for the mudra technique for years and years of time to practice, but definitely not possible to achieve the complete chakras by just doing for one hour a day. But according to me I believe if one goes into deep meditation state for years and years (minimum of 100 years) then the individual can obtain or unlock complete seven chakras to understand knowledge to lead life towards medicinal and scientific derivations.

A life without mudra is like an endless journey, because mudra practice activates the major chakra in human life. This chakra gives us the patience key to practice mudra for longer duration of time.

To understand the mudra and its technique to educate oneself then one must have a major key in their life as PATIENCE. In this Ghor kaliyug one hardly is having patience to achieve or to educate towards medicinal and scientific derivations.

Is this behavior of human in 20th century normal or its true about ancient people naming it as Ghor Kaliyug? I believe after practicing natya (dance) along with mudra, that it is 100% true that this human behavior of present generation is due to the generation of KALIYUG itself. One will never believe the importance of science until they practice the mudra in kaliyug generation.

A mudra is definitely a life saving, that's because one get education and the same one knows how to deal with problems in life through same education.

Life becomes beautiful after one practices perfect mudra technique to unblock one of the chakra in human life. A life is mystery if one does not unlock the chakra derivation through mudra.

As a dance scientist I have derived medicine and science through dance (natyashastra). When I was practicing dance for about 15 years almost, I was unaware that I use the mudra technique in my dance practice as well. This has bought me the major knowledge within me to spread it across the globe towards education and to derive both medicine and science.

It's ok if you are using the mudra technique through or while practicing dance (natya). But one must have the patience to educate their brain before they educate everyone in this world for their beneficial.

Life and its values are amazing if you consider yourself educating through ancient techniques of knowledge and perception. I believe our ancestors have left all the knowledge behind for our education purpose but as I said earlier that this is KALIYUG and nobody would believe it in fraction of seconds. It takes time to analyze things about the education derived by our ancestors with patience.

"Mudra is a dimensional practice towards Knowledge"

Fire/Agni

The most important factor in mudra is using source of energy in proper manner. This means that we have all the sources available to pray but the most important of all is Fire in achieving mudra knowledge.

We in India still make the prayers with Agni by lamping diya before the pooja (ceremony). It is followed since from many decades and the powers of Agni are as important as taking oxygen to live life.

Now to begin with any mudra the fire element is the most crucial factor to activate other elements like Water, earth, space and water.

Miracle Mudra

To activate each element in human life we need to hold along the fire finger with other elements. Now for example if one wants to experience or in need of water element in the body must practice Jal mudra (water mudra). This mudra is used when there is lack of water in the body, like dehydration and so on.

To maintain any element in the human body we need fire element to activate it. This means very clear that every element is been activated in human body through element of fire.

The above paragraph indicates that during pooja (ceremony) in Indian cultures speaks about purity while doing prayer. Even in mudra technique the fire element holds the activating channel to unblock other elements in human body like air, space, earth and water.

The element fire indicates purity in Indian culture. With the use of element fire technique most of the pooja (ceremony) in temple takes place.

It is even true that most of the people in India practice amazing rituals which brings peace and harmony in their life and in their brain functioning.

Now you might be thinking if the fire element is essential in activating other elements in mudra technique then how do we activate fire element in mudra knowledge?

Miracle Mudra

It is simple, other than thumb finger, close all the other finger and open only the thumb finger directing upwards like THUMBS UP, will activate the fire element in human body.

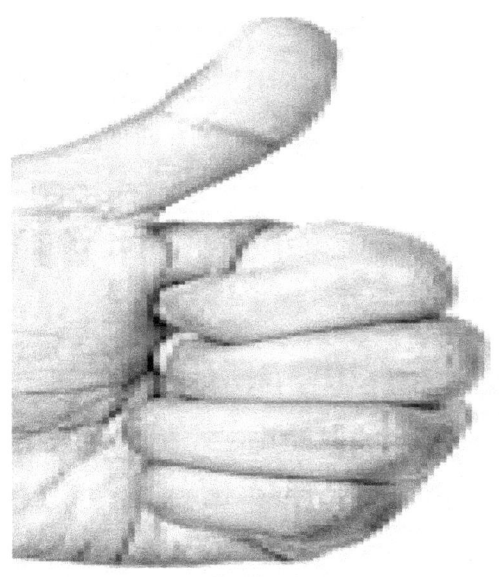

Miracle Mudra

With the help of fire finger one can activate other elements of energy sources in human body and in the universe.

Fire Mudra helps in retaining Agni (fire) in human body with respect to heat productions and its management. Imagine when you have lack of fire element in the body then the other elements slowly gets degraded and produces malfunctioning in metabolism in case of biochemistry.

Now many of the times my children has asked me one question frequently, that is how does the fire element plays important role in activating other elements. The answer is very clear without the

Agni (fire) one cannot produce heat in the body and without heat generation one will start falling ill from the air level to till water element.

Once the fire element is activated then other elements get the touch from the fire element. This is called holding mudra. Now while holding the mudra one must know that they must never press it with pressure. If this happens then while doing the mudra practice there might be damage causing to the nerves attached to all the other fingers or elements.

Once you start beginning the mudra practice, begin with YAVU mudra or Gyan mudra. In which the thumb

Miracle Mudra

finger (fire element) should come in contact with AIR (YAVU) element. As shown in the figure below.

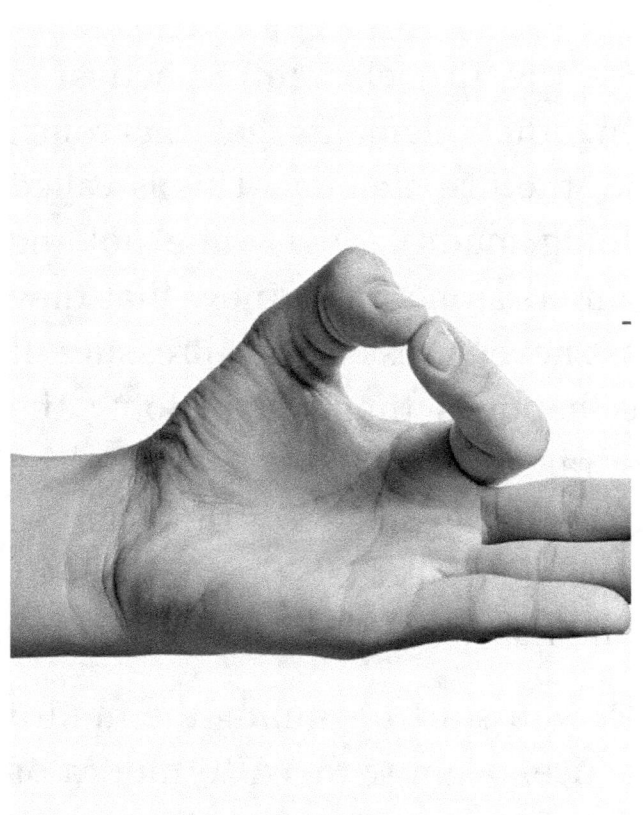

Miracle Mudra

The above mentioned mudra I (YAVU MUDRA) or in simple words AIR mudra indicates knowledge and its integrity. When the fire element comes in contact with Air element, the transition happens in through in terms of knowledge.

Fire mudra (thumbs up) is also the most amazing and powerful mudra to activate fire element in the body. Without the fire element it's hard to unlock the other elements in mudra technique o gaining knowledge and its perception.

A mudra feeling is wonderful if the person holds the mudra in perfect manner and not by pressing the finger frequently to hard way. This will lead to nerve damage in the finger which might bring many health issues like allergies or reaction in the human body.

In human body every element is important like water, air, earth and space. Even if there is lack of one element then it brings health issues which lasts for longer duration of time like allergies or reactions or stomach ache.

In very easy way I said earlier in this book that many Indians still now the moment they wake up in the morning they will do a prayer to SURYA (SUN) which is the major source of energy in planet earth.

The planet earth is covered with 75% of water just like human body which is covered with water element. So I believe as a dance scientist that both fire element and water element plays major role in understanding medicinal benefits from mudra knowledge.

YAVU/AIR

Air is the most important and is been considered as knowledge mudra or Gyan mudra for life. Knowledge is like air, which you cannot touch but you feel it within your brain functioning.

Air is as important as other elements in the human body to maintain chakra balance. As I said earlier Air (YAVU) mudra is considered as Gyan mudra for life and it's true that one who practices this mudra will gain tremendous knowledge.

To activate the air finger one needs to hold with thumb finger gently. The knowledge is invisible but one can feel it when you prove science through knowledge.

Miracle Mudra

Air is as important as oxygen, without oxygen one cannot live in this planet and one without knowledge will never survive in this planet.

Vayu indicates not just knowledge in terms of mudra, when you go into deeper study of mudra practice, you will be searching for answers in every particular mudra where how does this finger helps in particular way.

For example sometime there is mudra in which two fingers or elements or three fingers or elements are involved. Out of all three how this Vayu does is benefitting you.

Vayu plays an n important role in unlocking the major chakra in human body which is breathing technique you can say in normal way. A breath is carries not only oxygen n mudra practice but also it carries positive energy within the lungs and body.

Air or oxygen or positive energy, which one you inhale while you are in mudra meditation? I believe the one will inhale the all above three mentioned particles. But in a proper manner first the positive energy attracts the pure oxygen along with air inside the human body.

A mudra involving Vayu finger is definitely can be a MIRACLE MUDRA for some people who seek knowledge for lifetime. The Gyan mudra is been practiced in which Vayu and fire element are involved in been practiced by LORD SHIVA. Image is shared below for the activation of Vayu mudra.

The mudra technique gives the individual sense of understanding the energies surrounding. I believe mudra also gives a human brain complete enlightenment. We educate ourselves to be wise and stronger always but what happens if the brain is choosing a different direction while you life live? This is because we are living in 20th century and this is called GHOR KALIYUUG.

A wise man once told me that he has everything in his life at the age of 35 itself, but unable to complete my sleep during night time and having headache without any issues all the time. Mental issues were leading him to many health issues and later turned out to be man who has nothing remaining in his life other than end or death. When I introduced him to the miracle mudra, he was having trouble in sitting meditation with mudra for few months time. But later he got the answer and now he is able to life a better life without constant headache or any mental challenges.

Am not saying that what I found the miracle mudra is required or must be practiced by the every individual but am mentioning that miracle mudra I have found is precious and meant for me and for few more people who wants to enlighten their soul and life.

Considering vayu mudra is not just for knowledge but it also helps in stabilizing breathing issues or health problems relating to breathing problems.

To activate and to feel the mudra are two different processes. To activate one must produce fire in their body to activate fire element in the first stage and later activating the other elements like air, space, earth and water.

To feel the mudra power one must indulge into move-mental practice to seek patience key towards understanding perfect nature of mudra and their significances. A mudra through Vayu element leads to greater enlightenment if one practices for longer duration of time. As oxygen is necessary to live on this planet, just this we need to understand the breathing technique to reach out oxygen to complete human brain and body. Images shown below are various vayu related mudra for healing, education and to enlighten towards knowledge and its integrity.

One is called as RUDRA MUDRA which has healing powers as well as it relaxes brain functioning towards stability. In this mudra the Fire, Air and earth elements are involved.

Another image shown is Apana vayu mudra where Elements like fire, air, space and earth are involved.

Rudra Mudra

Apana Vayu mudra

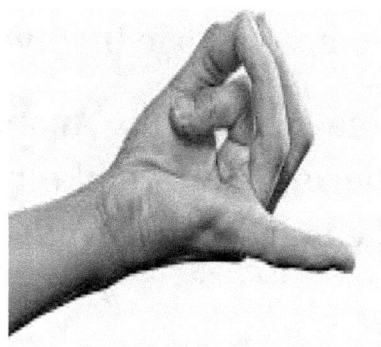

Space/Aakasha

Space or Aakasha element is the most untouched or practiced mudra in the world. The element of space is just like a brain, unexplored and unknown.

To activate space or Aakasha element in human body and brain one must hold the fire element along with space finger. This mudra of holding both fire and space element in common words call it Aakasha Mudra.

This element of space is just amazing if explored with patience to deal with self realization of human power to feel supernaturally.

Miracle Mudra

The one who activates this mudra of space will be flying in their thoughts and tend to move towards greater achievement than usual one.

IT is true that one who activates space element will believe in themselves towards greater enlightenment. The knowledge you receive through this activation of space will receive abundant knowledge in terms of science and its perception.

When I tried for the first time with fire element along with space element, I believed something great knowledge is residing within me towards medicine and science derivations.

Miracle Mudra

A life will have an empty feeling if one is not able explore the space and the universe towards knowledge. There is one mudra involving fire element and space element, which is called as KARANA MUDRA.

The mentioned mudra while holding the finger or elements of fire and space, the earth finger must bend. By doing this technique, one will have the control on their body and brain functioning peacefully.

It is true that KARANA mudra gives self confidence in understanding the power of your own and gives the capacity to control the power you have. When you can control your emotions and actions then one can enlighten their soul towards controlling powers and their limits. A life will be beautiful for those who practice this mudra towards controlling anger and gaining self confidence on every action you do throughout the day. The image shown below is KARANA MUDRA.

The space element activation brings awareness in oneself with respect to knowledge and its perception. The space looks beautiful with dark images surrounded with planets and gathered with many universes. As mentioned in previous line one will feel the energy of space after activation of space element in the human body.

One who practices the KARANA MUDRA will get complete knowledge over brain functioning through which one can control their emotions and actions. This way of using mudra gives one a sensual touch or feeling in the brain functioning with respect to war and peacefulness within the brain.

Miracle Mudra

Abundant space but blank it looks, just as I mentioned about Karana mudra in my previous lines once unlock the space element in your body, you will receive all the answers in the brain about how do you step and how step along with when to step for actions.

A life is very amazing one who experiences the space element through mudra technique. I believe when I was unable to control my actions and words through my tongue, I started using this KARANA mudra for self-control, through which today my actions are limited but perfect towards gaining and spreading knowledge.

Space element activation leads to opening up to another dimension of knowledge. It is believed that many of the saints and priests practiced mudra technique using space element.

In JAINISM I have seen the olden sculptures with space element mudra technique. It's hard to believe this technique has got mystical powers within and one will get great knowledge to store within the brain.

During practice hold the space element and focus on the space outside our planet. Am sure you will receive calmness within 25 minutes of your practice. As I said earlier that never press the finger but to hold the finger gently to seek the answer.

Earth/Prithvi

The most amazing and mesmerizing mudra in the world of practice is PRITHVI or earth mudra. This mudra to activate one needs to hold the fire finger align with earth finger.

The one who practices and activates earth mudra or PRITHVI mudra, will never have any health issues or problems related to health will be resolved by using Prithvi mudra.

One who has over weight can get reduction in weight or weight loss and one who do not have weight can get their weight back. Confusing isn't it. I believe this technique of holding the fire finger along with earth finger will give the individual immense power in terms of weight gain. While the mudra is activated it reduces all the health issues by activating not only earth element but also by activating the BHOOSPARSH (connection between legs and brain) or Connection of legs to the earth.

The who practices earth mudra will not only have weight gain but also helps in realizing self awareness and self control.

Medicinal values of this mudra are wide spread but along with medicine, it also helps the brain to restart the engine towards energy output. This means that one who practices earth or PRITHVI mudra will know how to spend the energy in sports or in games or in dancing or in some activity. By doing this excessive weight gain will be reduced frequently and soon the patient will realize that Prithvi mudra is made or practiced basically to deal with any kind of health issues or problems.

As a dance scientist when I first explored the Prithvi mudra, I had complete realization of how the medicinal effects are occurring within my body. To achieve this I had taken more than 6 months time to realize the medicinal values of Earth element finger or Prithvi mudra. Energy can be felt just within 10 minutes of this mudra technique practice.

Miracle Mudra

Image Shown below is the Prithvi mudra.

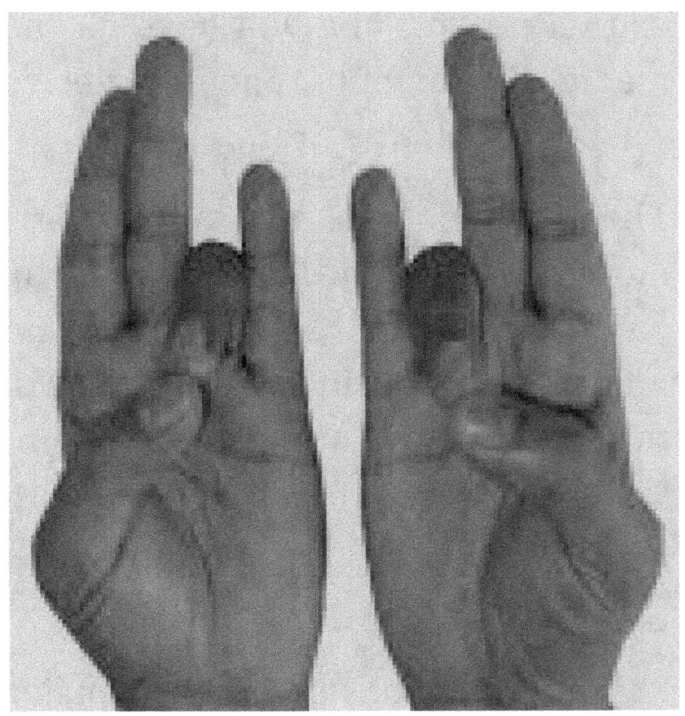

Earth has all the major components just like a life, which are water, air, space and fire. This means very clear that by practicing earth mudra or Prithvi mudra, one will have all the medicinal benefits and to derive science within the brain.

There is another mudra which is more powerful and has got amazing medicinal benefits than Prithvi mudra, which is PRANA MUDRA. In this mudra technique one has to hold the fire finger along with EARTH (PRITHVI) and WATER (JAL).

Prana mudra consists of three major elements, which are fire, earth and water. This mudra balances the complete body and energizes completely within the body serving against many diseases or disorders. This mudra is definitely for one who has health issues from long time.

Image shown below is the PRANA MUDRA, in which three elements together brings medicinal benefits within the body.

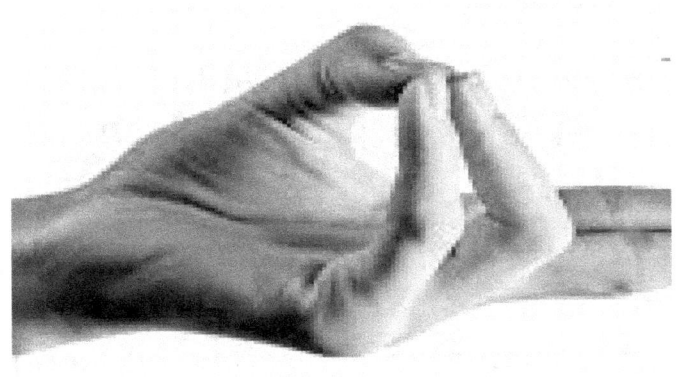

Prana mudra in English terms can be called as MUDRA FOR LIFE; this mudra has so many benefits that everyone or every individual can use this mudra for realization of medicinal benefits.

When I started this practice of mudra using PRANA mudra technique I was amazed to see such amazing results towards medicinal derivations.

For every health issue in the human body I suggest PRANA MUDRA as magic remedy. When I experienced this mudra of PRANA for the first time I had pain issues due to the dance (natya) in every day, but soon after practicing this mudra I do not feel any pain within myself anymore.

The elements of Mudra involved in Prana mudra are water, earth and fire. These three elements are the major necessary for regulating perfect biochemistry within the body.

There is many more mudra which has equal importance just like Prana mudra; they are APANA MUDRA and RUDRA MUDRA.

Apana mudra is so beneficial for every individual who has digestion problems. Soon after the food, one must practice the Apana mudra for complete digestion technique. Moreover Apana mudra activates many other chakras in human body.

Apana mudra involves elements are fire, space and earth. By doing this mudra after the food exercise, one will have complete balance with digestion technique in the biochemistry of human body.

Rudra mudra involves three elements in the mudra practice, which are fire, air and earth. This mudra as it states RUDRA relaxes the brain related problems like hyperactivity, anxiety and anger management.

After mudra practice for years today I believe that our ancestors have left us behind many techniques through which one can have complete blissful life to lead.

Element of earth is like complete answer for every human being towards knowledge of medicine and science.

Water/Jal

Water is the most essential part of life and has powers which can construct and demolish any other powers. The element of water is so essential in human body that every human is designed with 76% of water. This means very clear that we cannot survive without efficient water in the human body.

Activation of water element through fire element gives the individual experience about the liquidity in human body. A human being is completely dependent on three major aspects of elements, which are water, air ad earth. Life is covered with water to maintain perfect biochemistry functioning.

To activate Jal or water element in the body then touch fingers gently, which are last small finger along with fire finger. This mudra or technique has unique powers in maintaining water content in the body.

By keeping the body with proper water management the brain can have additional qualities like perception, goal, understanding and eager to learn more.

Usual problems associated with water are, depression, anger, anxiety, restlessness, stress problems, neurological disorders and many more.

All the above mentioned health issues can be dealt by using JAL MUDRA or water mudra. There are plenty of health issues which start from water imbalance, especially dehydration.

Water element along with fire element called as Jal mudra which can be a major solution for pimples on forehead or on face. The element of water is as essential as other elements in the human body, just like air, earth and space. But water plays major role because our body and the planet both are constructed with 70% water.

It is easy to do mudra or to hold the mudra with just a sense of touch, but hard to have patience to sit for longer duration of time, but through water element mudra one can even develop the patience power to sit for meditation for longer duration of time. This is because once the brain and complete body has the perfect water balance then the brain functioning will be greater than usual times.

The images shown below are various mudra techniques, using water element finger.

Miracle Mudra

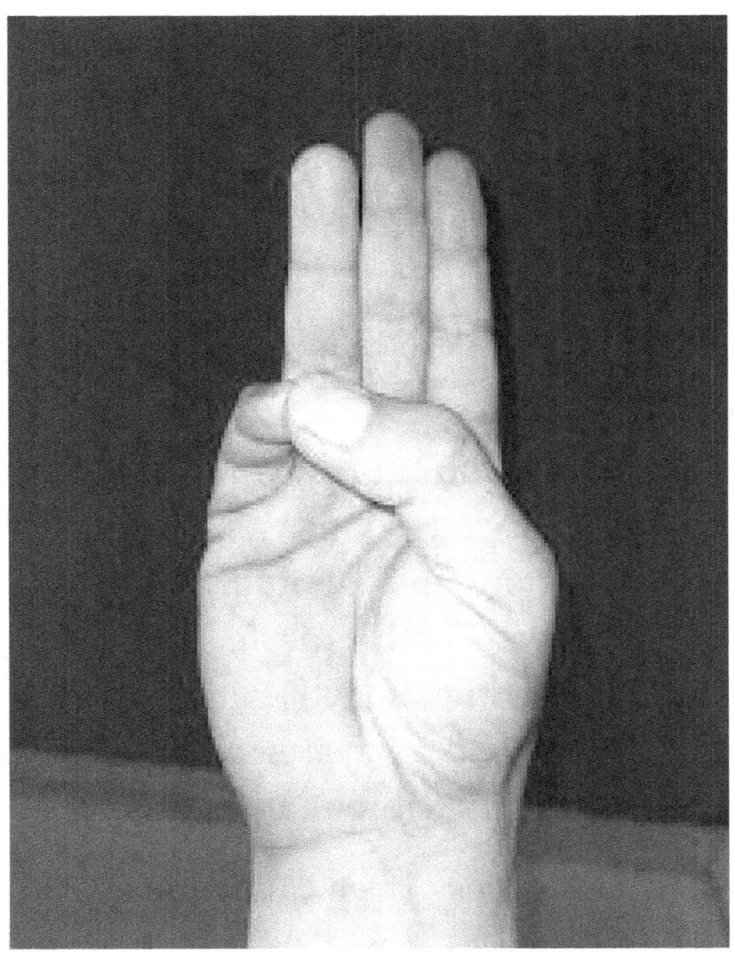

When I started the practice if water element, it was just within 20 days of time I have received many health benefits especially towards my skin and water balance in the body for biochemistry. Both element of fire and water brings peace in the brain functioning.

The one who desires of learning more knowledge and its integrity, one must practice Jal mudra to bring knowledge with perfect water balance in the body. Knowledge is also just like water; it flows automatically without any pressure.

To feel stress free and to again knowledge towards science derivations or towards service then people must practice Jal mudra or to activate water element in the body.

To maintain good kidney or to have to good skin qualities in life, one must practice Jal mudra or water mudra. Mudra with water element is like a magic spell for every individual towards gaining patience and true positive knowledge.

All the above mentioned information about all five elements, which are fire, Air, space, earth and water has all the equal importance.

One who believes to practice to activate only one element will have mixed emotions in life towards knowledge and its perception.

As a dance scientist I believe people who wants to try out he mudra technique in their life must start will with fire and sir in the beginning, which is called as Vayu Mudra. Later go ahead with all other three elements to explore. Activation of single element might also bring revolution in brain functioning. It is true that one must have intuition about which mudra to choose. Knowledge can be defined in both the ways according to me, one is positive way and other is negative way. There are people who seek benefits from mudra technique in very fast manner will never understand the importance and value of mudra towards knowledge and its perception.

Major Mudra

There are many mudra techniques available in this world, thanks to our ancestors towards spreading knowledge and passing the answers to us through mudra technique.

Indeed there might be many mudra techniques but there are few mudras which are called as major mudra according to me. Then which is the miracle mudra to consider the best off all, you might be thinking? As I said earlier that patience will give you every answer towards unlocking all the knowledge towards enlightenment. As a dance scientist I found out that there are seven mudra techniques to unlock all chakras in human body. In which I have practiced few till now in years. There are seven chakras in human life and each one of them has significant powers to activate all essential senses in human body, soul and brain.

I believe one must practice starting with normal mudras and later to adopt these techniques to get the specifically amazing results to gain positive energy to derive education and to derive medicine along with science.

The images shown below are major mudra, which are very powerful in nature towards gaining knowledge and its perception.

Miracle Mudra

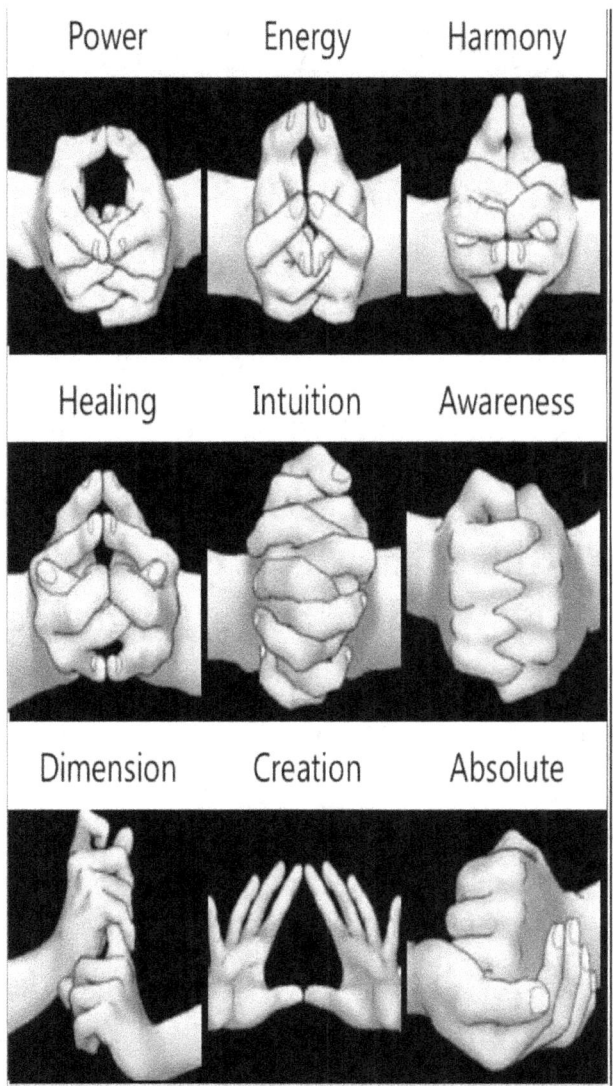

Human life and Mudra

It is true that human life has got many answers to explore as well as to experience. Many individual in the world still do not believe in their own hand powers, in fact many believes in others hand than their own hand.

It is true that we are living in KALIYUG where a flower might be illusion, fruit might be a poison and poison might become the remedy to live life.

Indeed the above line mentioned is 100% true because in kaliyug we believe in what everyone is showing to us. Reality is an illusion in the kaliyug and time is irrelevant.

Miracle Mudra

A mudra is as complicated as human body design. If you look into the human body, the direction of each nerve is designed so well that today we use this direction of nerves as MUDRA TECHNIQUE for education, healing and to derive science.

Our ancestors have left us behind with many knowledge and answers through many techniques like, mudra, natya (dance), Swara (sing) and words (shabda). Today even after knowing many amazing practice rituals we lack behind in health related issues due to the ignorance towards good knowledge or positive knowledge. This is all due to the generation we live GHOR KALIYUG.

Miracle Mudra

A mudra is derived and practiced by many of our ancestors, not only from India but also from the entire globe. If you could observe and see the old pictures or images or sculptures of every god and goddesses, there is mudra involved in their hand posture.

Human life is a blessed life to get the amazing fruits in life but we ignore it in such a way that the knowledge is considered as negative feelings or attraction towards gaining enlightenment. The negative knowledge in the present generation is been considered as the major knowledge today and all this due to the generation we live on GHOR KALIYUG.

Human life as I said earlier in my note that is a gifted one with many form of knowledge towards achieving medicine and science.

A life without medicine and science is like life without oxygen or life without direction in life to lead. It is also believed that many individuals get the answers after their ritual practices towards every obligations or queries asked by their own mind or brain.

I have received many answers along with the purpose of my life. This happened to me after practicing for years along with natyashastra (dance), mudra and then deep meditation.

Every time the knowledge can be felt and realized in each and every surrounding of our own nature but in this generation us hardly able to see it through naked eyes. Why is it so? Because I believe in this generation that knowledge is spread and used against positive energy through negative thoughts.

One can define the knowledge in many ways along with one cannot even see or read the knowledge in this generation. We are in a world today living with narrow minded people, just like everything is I WANT.

A mudra allows the brain to unlock its major portions or patience key towards understanding knowledge and having capacity to read it. It is definitely the first experience in mudra technique gives you or shows to certain path or direction.

According to me there will be a magnetic pull towards particular mudra after span of years to know your own miracle mudra. I have found out my miracle mudra after just after practicing for 12months regularly, each day with 2 hours minimum exercise with mudra technique.

How to use Mudra

To know your mudra to know how we take the benefit or significances from mudra technique then one must adopt the basic technique of starting the mudra practice by activating every element like, fire, air, space, earth and water.

The basic stage of activating the all elements includes the meditation for 15 to 30 minutes minimum for understanding the technique.

To use mudra technique or to learn the technique over learning holding finger perfectly for mudra, one must know that mudra chooses the person accordingly sometimes.

Miracle Mudra

Believe me, it's a magical energy when you hold the mudra perfectly without holding the finger tightly, but to hold it or give a touch in a gentle way.

Many of the times people who choose their own mudra or they select the mudra accordingly. I believe both the ways possible. Sometimes mudra chooses the persona and sometimes people choose the mudra.

When the mudra chooses you then trust me it's your miracle mudra, hard to achieve and hard to have patience while doing such mudra. Such mudra brings complete enlightenment to your brain functioning towards human life and its integrity. Very often seen that the people use or search the mudra for their fast recovery or to find out the fast answer? Every time you think this way while practicing mudra, it gives you or suggests you answer in whichever way you looking for, but most of the time due to this problem people easily go to another direction which is not built for them.

When I started the mudra practice particularly I found that I wanted to take mudra technique for enlightenment of my soul towards finding all my answers in a right manner. But if I could have practiced with ego mind to build my own universe then the mudra would have helped but might have taken my life to another direction in which I wasn't satisfied or be happy for what I gain. So I suggest most of the people to use the mudra technique with pure heart to seek the answers which brings happiness not only to your soul but also to every living particle to feel the energy towards positivity.

Knowledge through Mudra

How do define knowledge? Is it through reading or moving or stepping or by creating or by destructing or by energy? Any knowledge is important for me as a scientist, because there is existence of knowledge within every source of energy.

Most of the time we chose either moving energy or by adapting external energies to complete your actions like machines or we choose the energy which is within every human cell.

The one wants to gain the complete knowledge will have the prevention powers towards medicinal illness and will always make the movement accordingly. In this generation the mudra knowledge is considered to be the weakest language of knowledge, because we in this generation of GHOR KALIYUG would love to spread only negative knowledge than the positive one.

Knowledge derived from mudra technique is as beautiful as a blooming flower. It allows the brain to expand its direction and values energy surrounded by. Mudra explores the knowledge centre in brain and gives a potential energy to store more knowledge and more importantly the mudra allows the brain to recover or has remembering power of all the knowledge received till the age you received.

Today the present knowledge giving centers are just focusing on the words written by our ancestors and most of the individuals are having the feeling about the knowledge but do not want to experiment and experience it through practicing.

All the beginning of my life I considered dance as major knowledge but later starting realizing that there are much other energy surrounded by me and all the other energy has the equal importance. If am dancing or practicing dance I just forget the world and just focus on all the possibilities in making gracious movements with all the emotions attached.

As I mentioned earlier that one must practice the mudra technique to allow the brain to expand its vision as well as its dreams.

Knowledge is beautiful and sounds really great if one takes the knowledge and later expands it and spread it as knowledge to people around the globe to serve humanity.

Miracle Mudra

We all know that today we live in the place where we do not give importance to plants and tree. I believe they are our real ancestors in this world to give us medicine, knowledge, food, home and science.

Just like our tree and plants we are having another source of communicating with god through amazing practice of mudra and reaching the brain maximum capacity toward blissful life.

Mudra give knowledge towards exploring brain, it's give knowledge towards expansion of brain and gives knowledge toward understanding universe and its energies.

Miracle Mudra

Positive Energy through Mudra

There are two powers in this universe, one is positive and other is negative. It's proven than the atom has both these particles of energy to create an external energy.

Energy cannot be defined according to me, but can be felt, realized and expressed. Both the energies are required in this world to have a life, imagine there are only people with positive energy or negative energy to lead life. Is it possible that way? I don't think so.

Both the energies collide with each other to form atom. Atom represents or sparks both positive and negative energy. Now coming to human life or life on this planet, every human brain is covered with positive energy and negative energy.

So when there is existence of both the energies of positive and negative then why one must build a positive energy while leading life? Imagine the brain is telling you that apple is poisonous, so do not eat. But the other brain suggests you that this apple might be or definitely having life saving energy to survive, which one would you prefer or choose at the end? Positive right;

Miracle Mudra

Life has given us both the choices of choosing good or bad. It is often said and realized that the one will get the positive energy through Mudra practice. Not even a single time you will receive the negative impact on your brain. But what if the brain wants to use the mudra for serving negative energy? Simple, the mudra will serve you the energy but nothing last forever and KARMA comes into pay if one practices that way.

Do you want to use the mudra for healing then it will definitely serve as life saver but lifelong will you hold the mudra and request yourself to protect you from same place? No right. In this way it's better to make movements and then to achieve the mudra towards medicinal derivations as well as to serve the humanity.

I request the reader that if you have negative energy or feeling towards your knowledge then starts mudra practice as soon as possible to convert you negative feelings to positive note.

Mudra technique of knowledge not only gives the positive energy but also helps in dealing with negative feelings towards science derivation. Mudra and its knowledge are wonderful and effective if one uses for the benefit of others towards spreading positive energy to serve humanity.

Positive energy one can utilize to help other and not to save within themselves for their benefit. Did you understand reader with above lines? Explanation – if one tries to save all the positive energy to themselves only then the energy won't be to having much power to deal with negative issues. This is because if you create energy through moving or by mudra or by any other technique, one must spread it to the individual who are in need of positive energy to save humanity.

Positive energy is very precious but won't last for long time unless you spread it across to save people to save yourself. If you are able to pull the complete energy of positivity into your sol then use the energy to help others to help yourself to save it.

In my previous line I have shared my biggest knowledge or energy through which today am serving too many HIV infected children to help them come out from negative impacts or energy.

As a dance scientist if I could have saved all the energy I have gained through natya (dance), mudra and meditation then probably I would have been evaporated. Trust me I said the facts. I believe in sharing the positive energy to the one who needs it or searching for it.

Many even asked me that what if the person learns complete technique from me about natya (dance) mudra and meditation then they might use it for their beneficial only. Then what is the use of spreading to those who want to serve only themselves. I believe they are serving one or other person who are in need or search of positive energy. If not the energy what they took from me as knowledge and stored energy from their practice will never serve because there is no goal or aim to help humanity.

Dance and mudra together brings something greater energy to deal with many health issues and to derive major science.

Human Brian and Mudra

It's a miracle or human brain is designed to construct, create, preserve and destruct towards its own benefits.

For the first time when I was studying mudra I was unaware of the powers I received through it towards medicine and science derivation.

Human brain is best suitable for the mudra practice along with yogic postures to derive space within the brain and outside the space.

Miracle Mudra

Mudra technique or variations are so well studied till now that one can completely unlock the brain functioning towards many derivations like medicine and science. Complete variations of mudra techniques are wide spread in the Google or in the world through many gurus or teachers.

The human brain responds so well towards mudra technique that one can easily determine certain goals in life towards humanity. One who serves and spreads knowledge through mudra technique will be always happy with smile. Mudra and brain are two major miracles for me as a scientist.

There are several portions in human brain, once they are unlocked the brain feels energy and can see the complete brightness throughout the day or life.

The one who preserves the energy and serve their own brain functioning through mudra technique will be always in positive attraction and will be suitable position to understand energy and its integrity.

Once I was in complete shock when some portion of my brain functioning expanded and gave me answers within my own brain with sense of touch through mudra. This is when I started to write the MIRACLE MUDRA.

Miracle Mudra

Every mudra is miracle but sometimes I believe mudra chooses you or either you choose the mudra, answer is collected in brain with specific mudra to a specific person.

Mudra and brain are two major answers where both the energies speak to each other to find out the purpose in life to lead. What happens when you find out your purpose of living in this world? I believe you will smile to it and you will either look into your hand or in the sky for thanksgiving.

Energy in brain is stored in very well manner, in case of mudra technique the energy is not only saved in brain but it helps to expand its vision towards dreams and directions.

The energy produced during or after the mudra practice must be utilized towards medicinal derivations and to derive science by reaching out to people who is searching for such answers.

Every time you do the mudra practice you will gain the positive energy which has all the powers of educating your brain towards revolution.

What is the meaning of revolution? It doesn't mean that one will have all the powers just like that. One need to earn the power and with power comes responsibility to spread it the needy for saving humanity.

Brian is the major ultimatum. One who understand the brain and who trust in positive energy of god will always have powers within themselves to deal against any kind of negative knowledge or perceptions.

The mudra technique has expanded my views and vision in such a way that I have an answer today for every negative aspect of knowledge.

Believe and beliefs are two parts of brain functioning. One who believes in their own positive portions of feeling will always know how to deal with problems or issues related to any subject in the world. The same believe must be spread to children to educate that where there is negative aspect of learning knowledge, there has to positive aspect of knowledge to deal with it.

Beliefs are spread for both personal gain and for universal gain by spreading beliefs. One who spread the positive beliefs will always have the answer to deal with all the negative issues and one who spreads knowledge through negative aspects will never know the meaning of HELPING HANDS in their brain functioning.

I have met people who believe in spreading positive knowledge only in their life and they are called as GURUS or TEACHERS. Teacher who spreads negative knowledge will have negative impact in their life towards understanding knowledge and its perception. I always request the teachers to spread positive knowledge and trust me to be in this position as scientist; I have followed my guru in understanding only positive aspect of learning towards gaining knowledge.

No student is bad in this world but one teacher could bring bad; a positive note in the brain always gives you best knowledge to deal against any kind of situations in life.

Mudra a Major direction

In the beginning of my mudra practice I was completely into dance (natya) only. But soon after I started practicing dance along with mudra, I have received some miracle answers in brain towards knowledge.

A direction is very important in life in which the life chooses two of the possibilities. One is positive direction and other is negative direction. The positive direction in brain brings smile and spreads smile, but the negative one will always try to spread the negative across everywhere but very few people get affected through it. Most of the times the negative approach do not stand for much longer time, but the positive one will stay forever towards knowledge and its perception.

A direction is as important as every other decision in life. Every time we must ask ourselves that in which direction we are leading and what is the purpose of this direction. One who understands this will always have positive note or pathway in their life.

Imagine the life going in wrong direction and the brain never understood its path. This is happening to many people in this world who seek for bad knowledge or negative knowledge. The one who adopts mudra technique will always go towards good direction or finds the best possible direction in life towards gaining positive energy as knowledge.

When you start the mudra practice remember you must have appositive goal to achieve then later automatically every elements in your body will understand the situation and you will go ahead with best practice with mudra technique towards medicine and science derivations. Education must be involved with mudra technique in the present generation for the children who wants to seek tier direction in life for happiness.

What is direction by the way? I believe the direction is the major answer which every individual must seek for from the beginning of their life. for example if one individual is having a life without direction then trust me there are no images or dreams played in the brain. This is due to the lack of confidence and influence of negative energy in their brain towards knowledge.

> Every mudra gives you amazing directions in life with positive energy and thus by practicing mudra, one will not only get the medicinal benefit along with science derivation but also will get the positive direction in life to lead towards education or knowledge.

How to unlock the powers of mudra effectively

To unlock the powers of mudra technique and to gain its significances in life one must go deeper into knowledge perspective to achieve the goal.

After a year of practice with mudra I had many other people requesting me to share my mudra technique so that they will achieve the happiness and blissful life. I gave my knowledge and shared all my techniques to my friend, but my friend still didn't get the benefits of the mudra practice.

Why is this way happening then? This is because my friend wants the mudra to give him all the answers within few days of practice, which is definitely going to bring only chaos in his brain functioning about answers in the brain from mudra technique.

How to unlock the powers of mudra technique effectively? Simple answer is on who wants to seek the answers in a quick manner within few months of practice then one must adopt movements practice like dance(natya) along with mudra technique to achieve in faster rate towards knowledge gain in few months itself.

I practice dance in the beginning and then go towards mudra to complete package to gain knowledge and its perception. When I started the mudra technique alone I found out that it will take years for me to gain knowledge from this technique but when I tried mudra technique along with practice over movements like dance, I found that the technique works much more intensively towards gaining positive knowledge within first 20 minutes itself.

Amazing right? Instead of sitting for meditation for days or months or years long, one can achieve the best results of mudra technique by practicing the dance in the first stage and then later moving to mudra technique to complete task for gaining knowledge. The one who practices mudra technique without he movements will also get the benefits but after practicing for longer duration of time.

Is it good to practice mudra technique along with dance technique? (Smiles and answers) yes, in every Indian classical dance form the mudras are being used to perform or practice or to do a prayer or tapassya.

So it's better to dance directly right? Instead of going from dance later to mudra for meditation the direct technique is better one right? (Laughs) it's not. One who wants to seek the answers from the beginning of mudra practice will get the benefits after being practiced with dance (natya). This way of using both the dance and mudra while performing gives one a blissful life but it's hard to request everyone in this world to dance Indian classical dance itself through mudra technique to gain knowledge.

Every individual in this world has their own way of expressing through dance and I believe you dance accordingly with your own form or style but later adopt after practicing dance one must sit for meditation holding mudra for 1 hour minimum to achieve good results or to reach positive knowledge in brain.

Miracle Mudra

I know that my reader is waiting for the final answer about which mudra has the miracle powers to enlighten the brain completely towards education.

Before I announce the miracle mudra one must that if you are reading directly the final chapter of this book then you wouldn't be able to understand the concept and perception of mudra technique towards gaining knowledge.

Indeed one must read complete book to know the secret powers of each mudra towards knowledge and its perception.

Miracle Mudra

Before I call this a Miracle Mudra, I found out that this posture of mudra technique is the greatest symbol of education or symbol of powers existence in this world.

It's true and amazing when you practice the miracle mudra, one might get complete enlightenment and sense of all the energy within the body and soul.

This Mudra of miracle powers is called as UTTARABODHI by few people. But I call it as GOPURA or PYRAMID mudra.

Through this mudra I found out the major answers which my brain was looking for to enlighten towards knowledge and its perception. Image shown below is of MIRACLE MUDRA.

Miracle Mudra

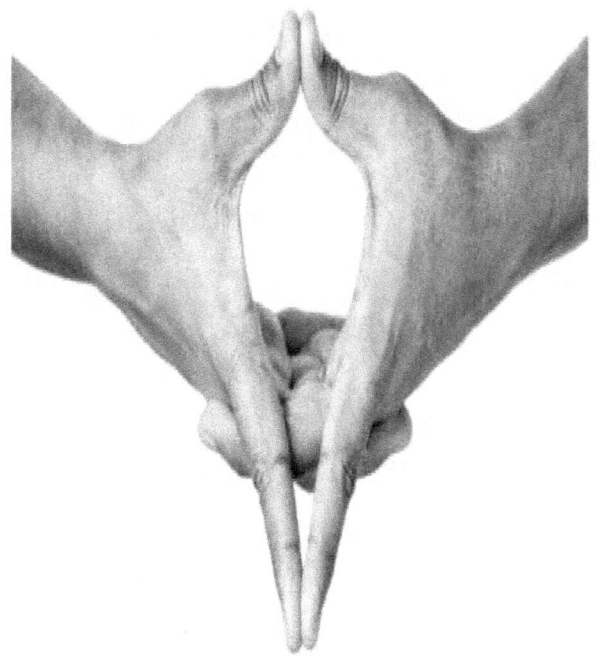

Miracle Mudra

Sometimes I call this miracle mudra as even RESTART ENGINE mudra. This is because when you try it perfectly in mudra technique of miracle mudra, you will gain the energy back in form like a child.

Engine will actually restart and begin with new life to lead towards enlightenment of brain towards knowledge and perception.

This mudra gives every individual a pathway or direction which they must follow to seek the happiness. When I practiced this mudra in the beginning I use to even call it as SLEEPING MUDRA. This is because in the beginning practice of this mudra one will achieve relaxation completely to bring sleep in human brain.

Indeed it is the miracle mudra because one will know the reason for their existence in this world. This means that one will find the direction to lead in this world towards knowledge and brain's enlightenment.

One who adopts this mudra will get all the answers to achieve in coming days. I call it as PYRAMID or GOPURA mudra as well. GOPURA means in Indian culture if you could observe every temple is built in this posture of mudra. So I believe there is a great power is present within this mudra.

Miracle Mudra

Miracle mudra gives one a bright light in the morning to see the sun rays clearly to feel the energy surrounded in this world.

There are various methods to hold this mudra perfectly for best practice. One way is to hold the mudra while sleeping, keeping the hand on the stomach and index finger (Air) element facing towards legs and fire elements are facing towards brain. This will induce or bring sleep in the beginning of the practice of this mudra for about 2 months but later one will achieve every image in their eyes towards gaining knowledge and to have enlightened brain towards living life in beautiful manner.

Miracle Mudra

There is another mudra I have found to have the miracle powers in human life. This mudra is named as KUBERA MUDRA but I suggest this mudra as CONFIDENCE mudra. Kubera mudra of confidence brings complete confidence in your life and brings complete energy in the brain and in the complete body towards making movements to achieve the goal. Kubera mudra must be practiced with sense of life to spread the humanity to achieve all the goals. One who consider or takes this mudra for granted will never get any miracle powers. But if you are using the miracle powers towards humanity then it will give you all the confidence in the world to achieve your goal.

With this last paragraph I will end this book but congratulations to a new beginning in your life towards positive energy and to enlighten the brain towards humanity.

Keep smiling and keep dancing and achieve the mudra powers after the dance practice for better results.

With love

Girish Kamanuri.

www.ingramcontent.com/pod-product-compliance
Lightning Source LLC
Chambersburg PA
CBHW060849220526
45466CB00003B/1306